TIME TRAVEL

A STORY OF TIME FOREVER

BY

Anthony J. Storino

ISBN: 1-4392-4085-X
ISBN-13: 9781439240854

Visit www.booksurge.com to order additional copies.

Dedication

I would like to dedicate this work to those of you who quest and thirst for knowledge. It is you who mankind will eventually seek out to find solutions for the problems of the future, and there are many. The world of today is becoming a scary place to live. The Cold War seems to be returning. But unlike in past decades where we had only one other superpower to deal with, today it is the third world countries that have the money gained from the oil or products they sell to the rest of the world that will be doing the harm. Using this money they can buy weapons and technology from any number of countries around the globe. The thought that a country can obtain a nuclear device—or, worse, some sort of germ that they could use to wipe out all animal life on the planet—compels me to write this book.

The human race seems to be a fifty-fifty mix of doers (the workers) and thinkers (the idea manufacturers). The visions and ideas that you have locked up in your brain which may someday release in the form of a finished product will, I hope, change the path of human evolution. Those ideas might even change the way the peoples of the world view one another so that in the future we can all be friends.

That is why building a time machine is important to me. The concept has been talked about and written about, movies have been made using a variety of ideas. We only need one good idea—the right one.

Who knows!
It may just be you who
solves the
code of
Time!

Contents

Illustrations

Introduction

Ever dreamed of stepping into a machine, pushing a few buttons, pulling a crank, turning a dial, setting the year you wish to visit, and—zap!—you're there?

Are you reading this book because you want to build a time machine, have a thirst for the unknown, or feel that you want to achieve something that no one has ever done before? That's great, because that is what it will take to be the first to accomplish time travel.

You will need to seek out those who, like yourself, want to find the answers. You must search through the dusty notes and journals of the forgotten dreamers who compiled all the data he or she could find on the subject of the past and present. Take what sounds like it could lead to a new thought, mull it over, change it, stir it around in your notebook, and see if it starts to look like a good Irish stew.

The Internet is full of information, but unfortunately it is also full of kooks! You need to sift through the garbage—and there is a lot of it flying around the airwaves. It will take time to do. Believe me, I have spent countless evenings pouring over Web sites, looking for the one thing that would click in my brain and light the idea bulb. Sometimes I did find a clue as it were, which led me to another and then another until I formed an idea with an answer. After all, if you just come away with questions and no answers, there will never be a finished product.

After what seems a lifetime of searching, I have decided that there are three ways to achieve time travel, two ways to achieve transfer to a parallel universe, and one way to place a person or object into a time warp. The latter could be of great benefit to those persons who are afflicted with an incurable disease. We will get into all of these possibilities later on in this book.

It is surprising to me how many people have never heard of Nikola Tesla or Kurt Godel. When I think of these two men, my imagination runs wild. The things these two men accomplished during their lives are unparalleled— from their first individual ideas or thoughts, using pencil on paper, and following their ideas to finished product. Although these two men lived in somewhat different eras, they were alive at the same time during parts of their lives. With Godel's math and Tesla's ideas from the visions he had, I believe if they had met and exchanged ideas, time travel would now be something in the past (pardon my pun). There is a good chance that there would be no need to write this book.

CHAPTER I
Is Time Travel Possible?

Theoretically, almost anything the human brain can think up can in someway be manufactured, built, manipulated, or reasoned to the final outcome or desire.

Ample proof of this can be found in our more recent history books. Around the year 1044, someone in China invented and developed gunpowder. They used it in the form of small fireworks that shot into the night sky and exploded over the heads of an invading army. When the news of this finally reached the shores of Europe in the 14 century, it was met with bewilderment and awe. What was this mysterious black substance that flashed and smoked. But then 150 years later someone like you or me living in Europe got an idea, a dream if you will, and invented the first guns and cannon. I wonder if you can realize what it must have sounded like telling someone that you could build a weapon that could mount a devastating blow to the enemy before he even came into view. It must have sounded like witchcraft to the people of the day. Throughout history many men have come up with ideas that seemed impossible, yet with imagination and foresight and the will to do, the impossible became the possible, and now in the twenty-first century it almost seems to be the norm.

On any given day, you can either read about something in a science fiction book or see it on a TV show—something about an idea, a thing that someone makes up for the plot or story. Then it seems like in just a few years, someone is working on a prototype of a similar idea.

Star Trek was a TV show just like that. You need to remember that *Star Trek's* first season started in the 1960s, long before cell phones, pagers, or wrist phones with cameras. But they used communicators to talk to one another anywhere they were located and at any time. Look at us today! Is it coincidence or foresight on someone's part? Just look around you and marvel at the technological advances that are being made every day, even as I write these words.

I can remember when I was in high school during the late 1950s and early 1960s. We thought we were a pretty advanced nation. Then came Sputnik. Unless you were living at that time, you can't imagine what it was like to find out the U.S.S.R. had orbited a satellite. That was Buck Rogers stuff to us. Most everyone's eyes were glued to their TV sets where the evening news would play and replay the strange and eerie sounds made by the Russian satellite as it passed overhead. My friends and I were in awe. Just think—now we could go into space, to the moon, to the planets and beyond. There was a manmade object orbiting the earth. Space travel would be possible, and it seemed to me at that moment, at that one and most beautiful moment in time, that nothing would be impossible anymore.

The 1950s-1960s was a time of product expansion and great ideas: the advent of color TV, hi-fi stereo, four-track and eight-track tapes. Then came cassette tapes, VCRs, and CDs—and none of these products would have been around if it were not for the invention of the transistor, which allowed electronics to be made smaller as it began to replace the glass tubes. Wires were replaced with printed circuits, printed circuits replaced by microchips, and the list goes on and on as man's knowledge increased. It's not confined just to the electronic category by any means. If memory serves, wax paper was the only thing available to wrap a lunchbox sandwich. The late 1950s saw the advent of Alcoa's aluminum foil. Wow! Something new, a kind of metal to wrap a sandwich in. Then came different cling wraps, and where would you be with out your good old microwave oven? Even nuclear power plants were just someone's daydream a mere forty-five years ago. I would venture to say that half of all the things you use in your daily life didn't exist forty years ago, and most were probably just dreams in someone's head.

Is time travel possible? You bet it is! But for now, I need you to just keep an open mind as we explore the realm and possibilities of time travel.

CHAPTER 2
What Is Time?

Time is just a word. A word that we Earth-bound humans use to describe when something took place. The terms yesterday, this afternoon, last week, month or year, one second ago, or three hours ago are just some of the words we use to describe when an event happened. Time is based on the Earth's rotation and its orbit around the sun. The Earth's rotation gives us our day and night. It gives us our hours and minutes, sunrises and sunsets. The Earth's orbit around the sun gives us what we call our year, 365.25 days.

As you can see, time is just a thing that man has invented, so that when we want to recall an event and relay that information to someone else, he or she can understand it. Everyone on this Earth, from an early age on, is familiar with the term hours, minutes, days, and years. It is a part of living and growing up on the Earth.

Without the Earth, however, you would have to invent a new way to keep track of events that take place. For instance, if you were to blast off this planet in a spaceship, you would no longer have the day and night. Therefore, you would no longer have the reference point of twenty-four hours to break up into hours and seconds. So what could you do? If you had thought to bring a digital clock aboard, you could set it at zero. As your trip progressed, you could refer to the clock for the elapsed time of the flight. One hour, fifty-one seconds into the flight, etc. If you had a large clock with a face on it like that of an automobile odometer, it would register about 999,999 hours. That's a long time, almost twenty years. At least you could register an event. This or that happened at 12,501. But the person to whom you were relating the message would have to go to the computer and find out what happened during that particular moment.

But what would happen if you forgot to bring a clock? Time as you and I know it would cease to be. No days, no hours, no years, no nothing, just now!

Can you imagine trying to tell someone about something that happened to you without the frame of reference known as time? You couldn't even tell someone when you were born, or how old you were.

I dropped a pencil! Oh really? When? I don't know! We rely a lot on this thing called time.

So what is time?

I believe time has always been and will always be. It never had a beginning, at least not one that we could understand, and it has no ending. I realize that this concept must be very hard to grasp, so for now just try to picture time as a ribbon or tape that started somewhere long, long ago. So long ago that we humans cannot fathom the length of it, so we say it always was and will always be. It was moving in the beginning and it is still moving in a certain direction. For our purposes, we will say that time is a ribbon traveling horizontally at a given rate of speed. I can't tell you the rate of speed because that notion would need miles per hour. See how hard it becomes with out the Earth?

Time is made up of a series of events. Each event is as separate as the grains of sand on a beach, or the stars in the night sky. Each event is broken down into equal individual segments. And each segment is separate, each is different, and each is in its own (for the sake of a better word) time slot.

Suppose you could make a calendar, with a space or square for each second or segment of time that has passed since the creation of the universe. In each square there would be an image, and by turning the pages of the calendar backwards, you could see the series of events unfold in reverse. Much like running a movie in reverse. In fact, let's pretend it is a movie.

If the Big Bang theory is correct, you could go back in time to when the Big Bang happened and see how the universe looked at that first instant. After that, if you continued running the film in reverse, a strange thing would happen. You would be looking at the oldest universe. Do you understand what I am implying here? If you went back into time beyond the beginning, which is the Big Bang, then what you would be seeing was the oldest universe before the last Big Bang. You would see all the species that ever developed in the universe, that had ever reached maturity; you'd see planets and stars die while new ones take their places. Slowly, as you reversed the movie of time, you would once again be at the year and place where you are at this instant.

I wonder if you would see yourself or someone else running the movie projector now? For, you see, as you continued to turn the pages in reverse, you would eventually find another Big Bang and then again at some point be at the instant you are at right now, again and again. Interesting, isn't it? If what I am saying holds true, then the universe is a giant ball with all the galaxies floating on the outside. I'm going to stop here, and come back to the round universe later.

Now to get back to where we were before I went off on this tangent.

Each frame of the movie film will have an impression, a segment of time. To be more precise, a segment of light captured forever on the film of time. By reversing the film, time seems to go backwards. Since time is not a film in a projector, you can't reverse it, but you can theoretically transverse the film of time in a time machine.

In today's world most people know what a VCR is. They also know what videotape is! How do we see the pictures on the screen since there are no individual frames on videotapes?

The picture or image is captured as an electronic impulse on the videotape. The VCR, with the aid of its computer chips, reads these electronic impulses and sends its findings through the cable connected to your TV. What you see on your TV screen is just a series of dots. Each dot represents a segment of the segment of time being viewed. As the tape runs through the

VCR, the VCR decodes the electronic impulses of each frame of time. The image or picture that is being viewed on your TV screen by the use of colored dots gives you the impression that the picture is moving, but it really isn't.

There is something that you probably do every day that is in some ways connected to time travel. Some of you get up in the morning and turn on your TV to catch the morning news before leaving for work. But how many of you ever think about what it is you are seeing?

It is again much like *Star Trek's* transporter, except instead of a person or thing being transported, the image is.

The television camera in the studio captures the image and breaks it down into atoms or electronic impulses. Then via a satellite dish the impulses are sent to a satellite in a synchronous orbit above the Earth.

(Synchronous orbit just means that the satellite is at an altitude and speed that matches the Earth's rotation, and so it seems to stay in one spot and doesn't move across the sky.)

After the signal is beamed to a satellite in space, the satellite relays the signal back to Earth over a very wide area. One satellite could conceivably cover an area almost as large as the United States. Your local TV stations or cable companies receive these signals and send them through the air or cable to your TV set. Then your TV set decodes these signals and voila! an image appears on your TV screen of the person or thing that was being viewed in the studios so far away. In this scenario it is your local newsperson.

All this takes place in less than a second. Therefore in today's world with a modern camera, a person can be thousands of miles away and you can talk to him or her as well as see their image as if you were talking to them in person.

Now, doesn't it seem strange that no one can transport an actual object anywhere? Why? We have the knowledge!

CHAPTER 3
Light Is Energy

You and I are made up of billions and billions of atoms, and the atoms are made up of electrons, protons, and neutrons. They in turn are made up of smaller particles called quarks, which in turn are probably made up of smaller particles that have not as yet been discovered. The point that I am trying to make is that they are all electronic impulses, or a better term would be energy.

Just as in the videotape, our bodies must be broken down into atoms to be transported to another place. Still, that's just being transported to another place in the same time! True! But what if you were to build a machine that would break your molecular structure down and cause you to halt in time?

The ribbon or videotape of time would keep moving and time would pass you by. You would in essence be going back in time, and if you stayed there long enough, you could when ready go into the future, the future that you missed while back in time.

To picture in your mind what I've just said, let's pretend that you are on a train. To be more precise, you are on the roof of the car just behind the engine. OK? Now, we will say that the train is moving at five miles per hour on a straight and level track, and that there is a wooden arch spanning the tracks with a rope ladder hanging down just ahead of the train.

As the train passes under this rope ladder you grab hold of it, put both feet on the first rung, and just stand on the ladder watching the train pass below you. As you stand there watching, you notice that each car represents a segment of time, like that frame on the old movie film.

You are now in your own little time machine. You just wait and watch time unfold. As you watch the train go by, you turn around and see the engine going farther into the future, a future that you have not yet seen. The longer you stay there on that arch, the further back into time you go, and the further into the future the train goes. When to let go of the ladder is up to you. You will land in the past somewhere in time.

There is a hitch in all this. The speed at which time travels through, I guess we could say eternity, is an unknown. Therefore, no one knows how long you would have to stay still to achieve time travel. It is my personal guess that time moves at different rates of speed, but we will talk about that later in the book.

This concept will get you into the past, but what about getting back to the present day or the future? Again, we will discuss this in a later chapter.

For now, try to grasp the idea that has just been laid before you. I know it is a concept that causes much doubt, but keep reading and let's see if we can change that doubt to a positive force.

People seem to have a hard time trying to fathom the idea that time is, shall we say, a thing, an entity unto itself.

Just as the planets, stars, space and we must also include the atom we cannot visually see are real, time is real!

For instance, we know that time is!

We also know that time bends and can be distorted. Also it is known that time can be stopped. Time can be warped. Time can probably be speeded up and slowed down also, but as yet to my knowledge no one has tried to figure that one out. So time is real!

Someday soon someone will gather all the data that has been recorded on time and a light bulb will turn on in his head, and he will say, WOW! It was here right in front of me all the time (pardon the expression, no pun intended). We have the knowledge now, but it will take a few dedicated people to bring about the desired result, which of course is time travel.

Some of those dedicated people might even live right in your own backyard, if you live in California that is. I read an article a number of years ago about the Aims Research Laboratory in Livermore, California, The article stated that Aims Livermore Lab was working on the development and building of a time machine. I don't know if the information was true or just fantasy but the article did say that the cover story was that they were building a super-laser, capable of destroying targets in orbit. Many years have past since I read that article and I believe they have finished the time machine and have tried it out. If they did not conceive a time machine then I think someone else did. I found out and it was quite by accident, that things are not always what they seem. If you search your mind for small events that have happened to you, or that you have read about, then later find out that what you recall is wrong, you may have been a victim of someone's journey back into time.

For instance, this has happened to me twice that I am aware of. I distinctly remember that James Earl Ray died. It seems to me I read it and also saw it on the news. I met a man last year who sounded a lot like James Earl Ray, and I told him so and that I thought that he could make money doing his voice on commercials. While in conversation I mentioned that it was too bad James Earl Ray had died, and a lady who was standing near by also confirmed the fact that he had died. We all got into a conversation and after awhile we said our good-byes and that was that. A couple of months later I thought I would look his name up on the Internet to find his final resting place. I do that with a lot of actors that I grew up watching on TV or the movies to see if they are still alive. Low and behold I find out he is not dead, but is staring in a Broadway stage show in New York City. I was in shock! When I found that article that said he was alive I couldn't believe my eyes. I pondered my recollections for many days after that trying to figure out how I had made the mistake and why, but couldn't find and answer.

The other event happened to me when I was working on my airplane model in the garage. I distinctly remember receiving a part I needed for the engine and when I went to install it, the part wasn't there. I checked the computer and found that the order had been placed but had not yet arrived as I remembered. And also I remembered picking it up at the post office. I hope you won't think that I am crazy or just writing this down to fill space. This is what happened to me and so now I must believe it. I have no proof other than what my mind remembers.

In light of this I feel that if not The Aims Research Lab, then someone else in the world has accomplished some sort of time travel and has distorted the space time continuum, causing small changes in our time. Not big enough to alert anyone, but very small events that most people just like me would shrug off as just a bad memory experience. Never the less you should search your own thoughts and see if there are a few things that just don't jive. It could be why we experience deja-vu. How many times in your life have you turned a corner or walked through a door way only to say to your self that I was here before. Or that I did this before even though you know that would have been impossible. The best explanation you can come up with is that it is a weird feeling. Maybe you have been there before and something has changed your memory. Changed it to a point that it is forgotten until something triggers that particular cell in your brains memory and it says to you, wait a minute I stored some of this before but that is all I can remember because the rest is wiped from existence. Maybe if you would like to, we can start a club where you can share your experience with others so we can get to the bottom of this. I hope that I am not the only person that this has happened to!

I've always believed that time travel was not dangerous and that you could go back into time only as an observer and that would be all. Now I am not sure anymore.

My convictions were that you could not change anything that has happened in the past. It has already happened and it is history. But lately I may be changing my ideas on this, mostly because of these unknown factors I've been experiencing. It could be that when time bends or warps, it changes the things around it for a short while, then when it returns back to its present time frame, some things don't all get put back in the exact same way. Maybe a few of the atoms get placed on a different line and so small changes take place. Kind of like looking at a photograph of a bunch of tiny flowers. Maybe one or two petals are misplaced on the picture. It would be so small that you would never see it, until maybe someday like me you recall something and it just doesn't seem right.

Well, for now it is just food for thought.

CHAPTER 4
Transporting Objects

I do not believe that transporting objects is as difficult as you might think. New ideas are popping up all the time. One idea is to use high-energy low-frequency waves. First used by Tesla during the early forties. I have mentioned him before in this book, and when it comes to electrical inventions, I think he must have been the best. He invented AC (alternating current) electricity that we use every day of our lives in our homes, offices, and everywhere. With the help of George Westinghouse as a money backer, they were awarded the contract to build the first AC hydroelectric power plant in the world, and they built it at Niagara Falls.

It is a very interesting story of how he came over to this country from Croatia to work for Thomas Edison. Edison was using DC (direct current) power plants for his electricity, but DC can only travel so far through a wire, and the farther it gets from the power source the weaker it gets. So Edison had to build power stations all over the city. Tesla's AC could travel unlimited distances from the power plant. After giving Tesla a job and working with him Edison found that he didn't like Tesla very much because of the way he discovered his inventions.

You see, Edison was brought up in the old school of thought: trial and error, using mathematical equations to solve his problems. Tesla, on the other hand, had visions of what he wanted to invent; he could see it clearly in his mind and then draw it out. He too used the trial-and-error method, but his methods were too far out for Edison to grasp, so he fired Tesla.

Tesla did work long enough for Edison to invent the only invention named after him, the Tesla Coil. But if you read any of the many books about the man, you will find that he also invented wireless radio, but Marconi got the credit using 17 of Tesla's patents. He invented neon lights, Electric submarines, remote controlled wireless torpedoes the list goes on and on. Yet this man is barely known to the general public. He died in 1943, penniless. What a shame!

There is a story—I don't know if it is true, but there are a few things that stick out as fact, however small. It is a story about the USS Eldridge, a ship of the United States Navy being used in an experiment that had to do with camouflage. There have been many articles written about it, a book and even a movie named the Philadelphia Project. Some articles claim that Einstein, Tesla, and even Howard Hughes were all part of the experiment. Supposedly the Eldredge was a brand-new ship anchored in the Philadelphia Harbor, and had not yet been commissioned for duty. She was sitting there idle so during the few weeks it would take to commission her the navy dept gave permission for her to be used for this experiment. They crammed the ship with all kinds of electronic apparatus that was supposed to, if the experiment was successful, render the ship invisible to radar. At the appropriate time, the switches were pulled and the ship became invisible. All that remained was a green-colored vapor in the shape of her hull. But then something happened and she really disappeared—only to reappear in another harbor many miles away, then disappear from that harbor and again reappear back in the Philadelphia Harbor.

It is said that when she reappeared in the Philadelphia Harbor, some of the crew looked as though they were on fire and glowing, and that some were caught melted into the bulkheads and decks. It is suggested that she then disappeared again with all hands aboard. The government has denied everything, and says she was commissioned and after the war was sold to   a Greek shipping company. There have also been programs aired on TV disproving that any of this ever happened, but there was a little tiny news article that appeared in the local newspaper from where the Eldredge supposedly reappeared the first time. It said something to the effect that three crew members of the Eldredge walked into a bar and had a drink, got into a fight, and left the same way they had entered: through the wall.

I saw a photograph of the newspaper article in a book once. It is an interesting story in that if it is true, that would mean that someone had achieved teleportation seventy-five years ago.

Theories and experiments on teleportation were very commonplace in the 1930s, 1940s, and even the early 1950s.

The British even took a stab at it, using a craft that would fly on electromagnetism. But it became apparent that without some sort of shielding, human beings could not survive. It seems that the electromagnetism would separate the iron in the blood and pull it to the edge of the walls of the blood veins, causing death. It is said that one of the pilots died.

There are those who believe that no apparatus is needed to transport objects. Take for instance the Rosacrutions, home-based in San Jose, California. They are an order that takes their beliefs from ancient Egyptian history. Within the body of the Rosacrutions is the Martinest Order. It is in this order that the members say they can levitate and transport objects. They are a worldwide secret society and do not boast of their achievements. They are one with nature and say they can do what has been said.

I guess we could scoff at that, but there are new studies being done at the present time concerning a parallel universe. A universe such as our own, a twin if you will, occupying the same space but in a different dimension. There would be another you and me in this other universe also, our twins! So it might not be so far-fetched as it sounds to transport an object to this other dimension. If a person could find the right frequency, an object might just disappear from our universe and appear in the other.

CHAPTER 5

Can a Human Survive Time Travel?

This is the question most often asked and the answer is filled with mystery and intrigue. Since man has never traveled through time, or at least I am not sure if anyone has traveled through time, we will have to answer this question using common sense.

Let's start by going back to the first automobiles. As the automobiles evolved the motors got bigger and stronger and higher and higher speeds were attained. The high speeds that a car could achieve became a worry to scientists of that era because they were talking about autos that reached around 60 miles per hour. Today we laugh at this, but to those who were living at that time, it was a frightful thing. There were those who said that a human being could not control a car traveling at such speeds, that his brain could not react fast enough to compensate in case of error or emergency. They may have been right when you consider all the traffic accidents we have today!

If you are old enough to recall the breaking of the sound barrier in a jet plane, then you may also remember that they said it could not be done. But as the test pilots flew their planes closer and closer to the speed of sound, it became apparent that maybe it could be done. The trouble was that no one knew what would happen if they did go faster than the speed of sound. Some thought that the plane would be destroyed that it would disintegrate in mid air. We know now that the sound barrier didn't destroy the plane and that the only apparent thing that happens is a loud noise like a clap of thunder, a sonic boom but with powerful shock waves that rattle homes below its flight path as the plane passes overhead. And not just one sonic boom, but many over and over again as the plane travels to its destination. That is why laws were passed to the effect that no plane can go faster then the speed of sound over populated areas. but the underlying point is: We just didn't know what would happen. In fact it was not until 2008 that NASA captured the first pictures of a jet plane the instant it broke the sound barrier. The photo resembles a jet flying through a white halo type mist.

In the 1960s NASA was doing space research during the Gemini missions and opened the space capsule door and took their first space walk. They didn't really know if the space suit would work. The astronaut could have been fried, but the suit did work and he returned to Earth well and fine.

In a previous chapter I stated that almost nothing is impossible and that anything we want to do we can do. I really believe that. There will be risks to be sure, and some will probably die before success is achieved. But mankind has this thirst and for the most part will achieve the goals he sets for himself. It might take many generations to realize some of his goals, but in the end, he will accomplish his tasks.

OK! Now let's get down to time machines. As you probably already know, most successes are due to previous failures or mistakes, and we build up our knowledge from those mistakes, correcting and refining until success is realized. The time machine that I am presently designing uses for the most part high-energy magnetic coils. But I am very much aware of what happened to the British when they tried to build their flying machine. There are new alloys in the world today that didn't exist when they tried to build their machine. Which, by the way, goes along with what I said in the introduction of this book. Many old ideas and inventions may not have worked way back when, but in our time with our new technology and knowledge, our modern computers and our abilities to form new compounds, a new metal, or plastic, who knows what can be achieved.

You remember I mentioned Nikola Tesla early on in this book. He has many, many patents to his credit. One is a lightbulb that, when refitted with new materials that have been invented during the last decade, will be brighter and cheaper to run, and will last ten times as long as the bulbs we are now using. I read an article not too long ago that a company had acquired the patent and are going to produce these bulbs in the not so distant future. That's why I always say research those old forgotten notes, go to the second-hand stores and look for old books that deal with electromagnetism and inventions of the past.

Find out for yourself what these dedicated inventors were trying to do with their ideas; maybe you can use their information in our time.

If I remember right, the heading of this chapter was "can a human survive time travel." OK! Let's explore this thought. Maybe we should start by asking what are the different ways in which time travel can be accomplished?

If we could get to a black hole, theoretically, time travel is possible! What is a black hole? I will try to explain it so that you may understand it better.

Imagine, if you will, a star, kind of like our own sun, with one exception: our sun is stable. Our sun isn't erratic. It is burning itself up at a constant rate of speed. But the most important thing is that our sun is not a hot star, as stars go.

Our sun is not hot enough to burn its final elements, so it burns its elements one at a time. Picture if you will a campfire. You start with paper and when you run out of paper you switch to small twigs, then to sticks, then logs. When the logs are all gone maybe you switch to coal, and so on until you have nothing left to burn.

So it is with our sun. When one element gets used up, the sun starts burning another, and so on and so on until one day about six billion years from now, it will have used up all its fuel.

During the final stages of its life, the sun will expand to engulf the four inner planets—which, by the way, includes our own Earth—killing all life forever. Then it will collapse in upon itself and explode, sending gases and material into space. The remnants of the material will condense once again over a period of millions of years and our sun will become what astronomers call a black dwarf. It will be about the size of our Earth, will cool and become a cold rock in space. Not a very nice ending for such a wondrous thing that created so much life here on Earth.

Now when a star is gigantic, about a thousand times larger than our sun and a lot hotter than our sun, it can burn up all of its elements very fast. Now, when that star uses up its fuel, the core gets hotter and hotter until you have a huge explosion. That is called a supernova. But because of the star's original mass, when it collapses in upon itself, the gravitational pull is so great that a teaspoon of its matter would weigh one billion tons. As it keeps on collapsing further and further, its core becomes a mass of protons and electrons. Because of its tremendous gravitational pull, the protons and electrons are then changed into neutrons, so that the whole core is now made up of nothing but neutrons. Thus it is now a neutron star.

But if the star were big enough to start out with, the gravitational pull after its collapse would be so great that even the forces that hold atomic nuclei together would be overcome, and the result would be a black hole. The gravitational pull is so great in a black hole that not even light can escape it. When it reaches this state, because of the gravitational pull it has, it starts to bend and warp the space around it. It sort of resembles two giant black funnels with the large open ends opposite each other. There are many black holes in our own Milky Way Galaxy.

So as far as light years and time travel are concerned, you wouldn't have to travel outside of our own galaxy to find one.

No one knows what would happen to a human being if he or she would enter the vortex of a black hole, but let's assume we are in a spaceship and we fly to close to it and get pulled inside. Of course we cannot escape it, so the first thing that would happen is we would start to gain speed. Since light cannot escape the vortex of a black hole, we must assume that we would achieve a speed faster then that of light. Most physicists would tell you that you can't survive and that you would be crushed and torn apart.

I, however, believe that if you were in your spaceship sitting at your controls, as you went faster and faster and eventually reached the speed of light, you, your ship, and anything else within your ship would turn into energy like *Star Trek's* transporter. Once in this state, you would be oblivious to what is going on around you and in that way protected against the certain death that physicists predict. Also, you have to remember that as you approach the speed of light, time will slow down and eventually stop. As you go faster than the speed of light, time will go backwards as you pass through the black hole and out the other side. At least I assume that there is another side! Once through, you would again start to slow down and your energy or atoms would start to reform until you were once again just as you were before you entered the black hole. The other equally amazing part is that this would probably have happened in a blink of an eye. The journey I have just described seems very plausible to me, but where would you come out? Your guess is as good as mine at this point. You might find yourself in the past, or the future. You might find that you are in a parallel universe, I don't know!

There are also other notions about going into black holes. Some theories say that there might be wormholes within the vortex along its walls, so to speak. And that these wormholes would act like doors to other universes, other times past and future, also parallel universes. Kind of like a detour to whatever, maybe even back to where you were before you went into the black hole. All that might be possible if you could maneuver your spacecraft into one of the wormholes. Or maybe the vortex has eddies and currents that would automatically pull you through one.

Another scenario! After you passed the speed of light and are just energy, what if you could still see each other and talk to one another? Your ship would still be intact and the controls at your disposal. You could steer it into a wormhole or any other place that happened to be there. We won't know until someone tries it, how exciting it would be.

Another way to achieve time travel would be to build a machine like the one in Chapter 9. Of course I won't ever know if it is going to work, because it will cost me too much money to build. It's out of my reach, but maybe not yours!

The third way to achieve time travel is one that I haven't completely ruled out. It has to do with using your mind. In this scenario you would build a machine that looks like a narrow bed with a round canopy at one end. You would lie down on this machine with your head under the canopy. There would be electrodes connected to your scalp over the parts of your brain that would send out the signals. I haven't really thought out the mechanics of it, but I did envision a drawing of what it would look like. It would somehow utilize your own electrical brain waves, boosting them to a great levels, and you would send out these magnified electrical brain waves that were in sync with the Earth's own electrical waves. I believe it is 7.68 cycles per second. Out side there would be a huge satellite dish that could be directed to any part of the sky. You would use your will and venture far out into deep space. Your body would stay right where it was, never leaving the machine, but your mind's essence would see everything just as if you were really there and communicate and converse with whomever you came into contact with.

Like I said, I'm not a scientist and I only have ideas and a yearning to make it happen, because you never know.

As I end this chapter I would like to say that as I see it, the human body will endure time travel because man will think up ways to survive it. Just as I have introduced you to these thoughts of mine, so others will build on your thoughts and so on and so on.

CHAPTER 6
The Fabric And Bending Of Space

What is meant by the fabric of space? Well, long ago it was thought that the matter that makes up our universe, such as our solar system, stars, the galaxies—we'll call it bright matter—was all there was to the universe and that all the black space was an empty void consisting of nothing but a vacuum. In recent years there have been a number of small but significant discoveries that we now take for granted. Knowledge is ever forthcoming, and the more you study something the more you will learn about it. It is the same with the universe. But until man could free himself from the shackles of the Earth and explore outer space, all his ideas and thoughts were just that, thoughts and ideas. Nothing more than, notebooks filled with science fiction that was based on all the knowledge that we could ascertain from what we thought we knew. History has shown us that science fiction has a way of becoming reality, and much to our surprise, we found out that there was a lot more to the universe than we had previously thought.

We found out, for instance, that there is dark matter out there so thick that it is blocking our view of what's behind it. We also found that when light from a distant star passes close to a large body in space, it dips and curves as it passes. "Light always travels in a straight line;" remember that saying? So what makes it deviate from its course through space? The only thing that makes any sense at all is the space surrounding solid bodies in the universe is bent or warped. If you could take the universe and flatten it out like a tablecloth, you could see how the bending of space occurs. Suppose you were to take a tablecloth and stretch it tight between four posts, and then place metal balls of different sizes on it a few feet apart; you would notice that the weight of the balls have made an indent in the tablecloth. The heavier the ball, the bigger the indent. Our little mock-up is a somewhat accurate visual concept of what's happening in the universe. All matter in the universe is occupying a piece of space and is producing a depression around itself relative to its size, weight, and gravitational pull. Unlike our blanket, where the space only at the bottom is affected by the ball, in the real universe all the space that surrounds a planet or object is bent—top, bottom, and sides.

The space around each planet, star, galaxy and so on, is being bent, or a better word for it would be warped. This warping of space is what alters the path of light as it travels through the universe. As the light passes close to a planet or object, it follows the contour of the fabric of the bent space, thus deviating from its true course.

The light also speeds up or slows down as it traverses ridges and valleys in space. So when someone talks about the speed of light, it is no longer a constant, is it?

Someone once said, and I believe it was Einstein, but don't quote me, that if you could shoot a constant beam of light into space, eventually it would return to the place of origin. How could it?

If you think that the universe is flat, just as they thought the world was flat a few centuries ago, then no, it probably wouldn't return. If on, the other hand, you believe as I do that the universe is round like a giant ball, then you can see that the light would return to its point of origin.

A round universe?

Yes, why not? Just look around and what do you see? From the smallest particles known to man, to the largest planets, to the sun and the super-large suns: Everything in the universe is round. Look at the galaxies that spin, and in between all these bright objects, in the blackness of space, the dust of creation. There is dark matter out there, and it covers the entire round universe, much like the oceans cover the Earth.

A few years ago, astronomers were in a state of shock because they discovered, with the help of the newly operational space telescopes, the galaxies at the furthest reaches of the universe are speeding up instead of slowing down.

I'm sure that most of you have heard of the Big Bang theory. It is the theory of how the universe began. It posits that all the matter in the universe was just lazing around floating, and that somehow it started to compress and started to heat up. And as it kept on compressing, it got hotter and hotter until it finally couldn't be contained and there was this huge explosion which sent hot matter streaming away at great speed in all directions. Later, as the matter started to cool, it started to condense and form the galaxies, stars, and planets.

I feel that this theory is flawed because it doesn't tell you where all the matter came from in the first place? I don't know the answer to that question either but the bible states that God created the heavens and earth and said let there be light and the big bang happened. I'm ok with that! Something or someone had to put the matter there so that it could explode. I as a human being do not have the ability to understand how the universe was formed, but what followed time and again I feel I do know.

Let's suppose that the universe is round like a ball, and that all the matter is moving away from one central point of origin, which it is! Then it stands to reason that all matter will eventually meet at the opposite point from where it all started, or the Big Bang. In this case, the matter of the universe will converge at one location because as the galaxies and all the other matter get closer to each other, the gravitational pull will get greater and greater and there will be a gigantic crash and explosion. Another big bang that will send matter back out into the universe to begin the process of beginning new life all over again, and it will happen again and again. Never-ending big bangs!

According to the theory of the Big Bang as it is taught now, the galaxies will eventually slow down because of the gravitational forces they exert upon one another, as if they were connected by an invisible rubber band stretched to its limit. They will slow down and stop dead in space, then they will start to move in reverse until they all meet at a central point and we have another big bang. I grew up believing this theory because it made sense to me at the time. But now with computers and space telescopes and all the new data, I can see it has a flaw. It is now known that the farthest galaxies are speeding up. They are most likely being pulled by other gravitational forces, such as nebulae and galaxies that we cannot see because the universe is round, and they are below the horizon line. Those nebulae and galaxies are on the other side of the universal ball beyond our line of site; we can't see them, but they are there.

Now suppose that I am right and the universe is a round ball; it could make time travel more accessible and easier to accomplish. Using our own Earth as an example with its continents and islands substituting as galaxies and planets, and the oceans substituting for the dark matter of space, we can readily see how this can be accomplished.

We have machines that fly and others that sail on the oceans, and those that can dive into the murky depths below. Using this analogy, let's explore what possibilities exist using our round universe scenario. (See diagram next page.)

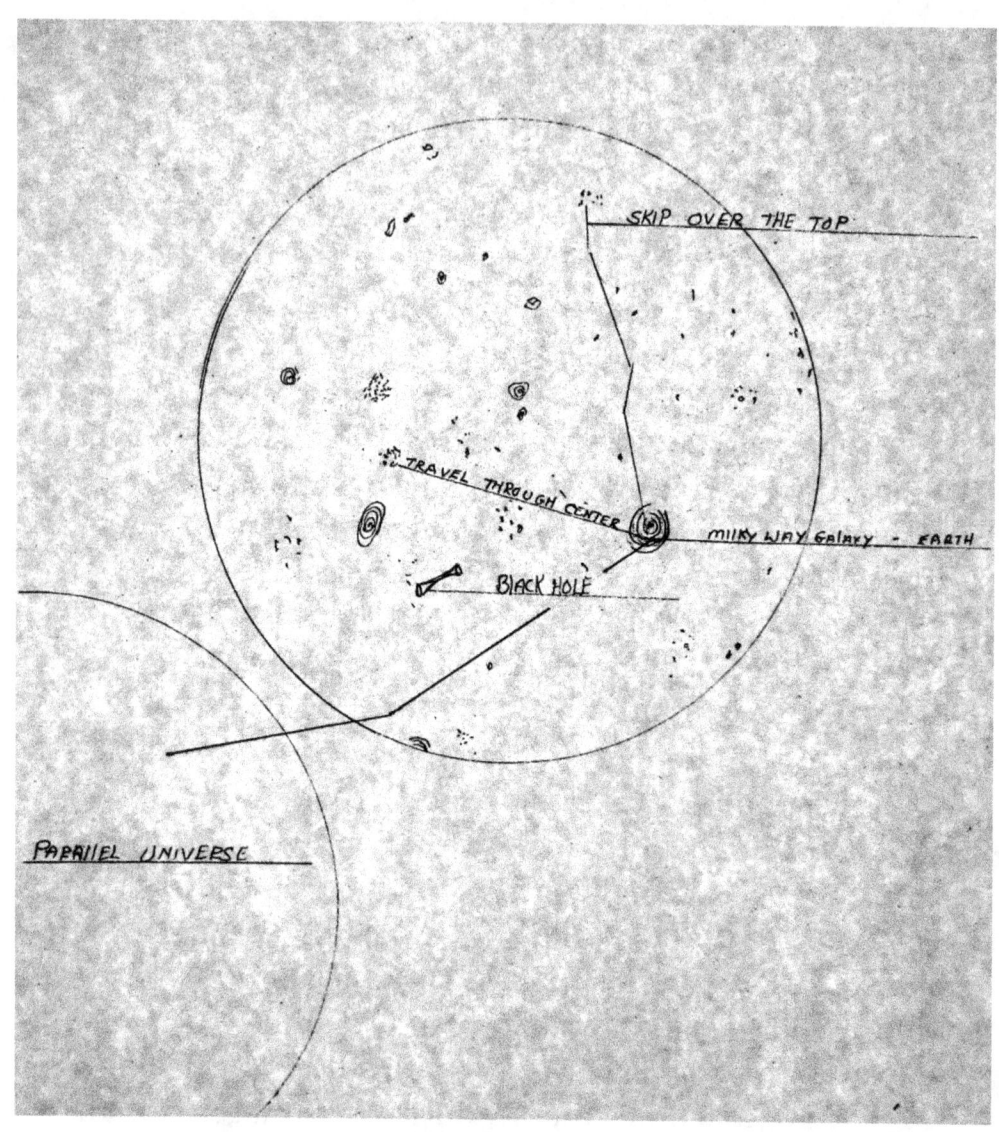

As you can see from our diagram of the universe, it may be possible to transverse around and through it.

For example, you could always do what we do now: blast off in a spaceship and travel the cosmos much like we do flying around our own world. Or perhaps you could shorten your trip by plowing right through the middle like a submarine. Maybe even try a skipping effect, like when you throw a stone across the water. It may be possible to skip off this universe and land on another one. This idea may sound strange but if we adopt the round universe theory, then it stands to reason that there would be many more round universes out there. You could transverse the void of space in the opposite direction or at right angles and reach the stars and galaxies in a shorter time frame. Of course a light year is a light year, and it doesn't seem that we are going to get to any galaxy in the very near future, unless you build a machine that can match the molecular bond and resonance of the universe. Sounds something like a time machine doesn't it? It is very similar to a time machine, but it would act as a transporting device.

To give you an example, let's suppose that point A is San Francisco, and point B is San Jose, California.

For our purposes we will say that it is exactly 60 miles from point A to point B. At 120 miles per hour, it should take you 30 minutes to cover the distance from point A to point B, correct? If we double the speed to 240 miles per hour, then double it again to 480 miles per hour, we could cut our travel time to 7 1/2 minutes. Lets double it again to 960, to 1,920, and again to 3,840 MPH; how long would it take us to get from San Jose to San Francisco now? It would take 93 seconds. If you were to double the speed five more times, you would shrink that time down to less than one second. Why not build a machine that could catapult you to point B in the blink of an eye? There are those who would say that the human body would be crushed and that the traveler would die. Of course that probably would happen if you started out slowly and stopped abruptly. That sudden stop from such a high rate of speed would most certainly kill a human if it were to occur using conventional modes of transport. But what if you were to build a different type of machine? Consider if I may, a tube machine, stretching from San Francisco to San Jose. Using electromagnetic power, you could design a unit that would fit inside the tube; the person sits or lies down inside the unit. Because it is powered by electromagnetism, no part of the unit would touch the walls of the tube. It would float and could achieve the speeds necessary to get you there in that blink of an eye. It would be instantaneous. You wouldn't feel the trip at all. One second you're in San Francisco; the next you're in San Jose. Because electromagnetism is pure energy, it might be possible to reach San Jose before you left San Francisco! But let's not go there yet—that would mean we invented a time machine.

Now some of you are probably thinking about what you just read and are also thinking that it won't work. Please keep an open mind and let me tell you why I feel it will work.

First: If all things everywhere are made up of atoms, then it holds true that everything is connected to everything else in the universe. At the beginning of this chapter, you'll remember I said that when you look up into the sky night to see the stars, there is dark matter up there also. All the darkness that you see is dark matter. It is not just empty nothingness. On the contrary, it is loaded with atoms and complex molecules. Some parts of the sky have more molecules than other parts, so the dark matter is thicker and you can't see through it. The same holds true but just the opposite: fewer molecules are present where you can clearly see objects far out in space. The chair you are sitting in right now is nothing but atoms. The house you live in and all that's inside it are just atoms. Some atoms are closely packed together to form solid objects, while others are loosely packed and we can walk through them. Air, and water for instance both let you push your way through it. Water sort of moves aside as you go. Add additional atoms to water such as what is in salt, and you can sit on it without sinking. It will hold you up.

What about glass? Did you know that glass is not a solid but a super-cooled liquid? If you were to place a glass window pane on edge for a long period of time, eventually it would be just a puddle. It would still be hard to the touch but a flat puddle on the ground. You can prove this to yourself if you have any very old buildings near where you live. If you look at the windows in these buildings close up, some will be thicker on the bottom than at the top, and some will have streaks in them where you can see how the glass has melted or flowed a little.

Since you and I are made up of atoms, and the air, water, and everything else known to man is made up of atoms, we are connected.

Being connected in this way makes it possible for you to pass through objects that seem to be solid. Have you ever seen objects that have passed through other solid objects without damage to either? One example that comes to mind is a photograph I saw once, of a piece of straw that was caught in the awesome force of wind during a tornado. When the tornado had moved on they found that a piece of straw had pierced a one-inch-thick board and was protruding through both sides of the board without any damage to the board or the straw. How was this accomplished, since both are considered solids? The tremendous speed of the winds of the tornado enabled the straw to pierce the board, but more importantly, the frequency or resonance of the atoms in the straw and board were matched at that precise instant. Matching the frequency is the important thing, not the speed. Therefore, if your body is tuned to the right frequency, you can pass right through what we think of as solid objects and maybe even into parallel universes.

Our electromagnetic transport would do the same thing, and if we find the right frequency no harm will come to us because we are just moving our atoms through other atoms, and in the future it will probably be as commonplace as getting on a bus today.

CHAPTER 7
The Space-Time Continuum

Thus far in this book, I have tried to convey my ideas that time is a real thing. It is something across which you can traverse and roam. OK! Let us assume that you built your time machine and that you do succeed in traversing to a place in the past. You're there!

Can you change the past and alter the future?

I don't think so. I feel that everything in the past is just a shadow of history. You can explore it, watch it, and enjoy it, but you can't change it.

Whenever the subject of time travel comes up, someone recites the question of what if you went back in time and killed yourself. How could you build the time machine to go back in time if you didn't exist? A similar scenario is if you went back into time and killed your parents before you were born. These only add more truth to my original statement at the beginning of this book, and that is you cannot change what happened in the past.

What about my memories of things that I felt I knew had happened and now I find out that I was wrong about them? I said that I thought someone had built a time machine and gone back into the past and changed something, and that incident changed little things here in the present. Does this contradict what was written in the previous paragraph? Let's explore my statement. I said that I did not believe that you can change anything in the past because they are just shadows of the things that have happened. I do believe this to be true, but we are dealing with atoms here and also atoms in the past.

I believe that if you built a time machine and went into the past to change something specifically, like trying to stop the John F. Kennedy assassination or warn people of a disaster or just try to change anything at all, you would fail.

You cannot purposely change anything that has happened, but I do think that maybe when you are traveling through time you can disturb the fabric of space and move some of the atoms and cause something to change in the future.

You probably already know, for instance, that no two pieces of matter can occupy the same space. Even though you are traveling in the past, the past has its own atoms and there may not be any room for your atoms to move around. Like I said in the previous chapter we are all connected to each other via atoms. If we go into the past, then we are taking an extra set of atoms into the past also—namely, us!

So now there will be two of you in the past. It was full before you got there. Something has to give way to let you in or for you to move about while there, and I feel that is what has happened to me as far as my memories are concerned.

Someone may have built a time machine and while in the past, as he or she moved through it, some of the atoms in the past did not come back together as they originally were. Let me see if I can illustrate this.

Suppose you had a box of sand, and some of the grains of sand were painted yellow, some blue, and others red. If you poured the sand into a glass jar and looked at it from the top, it would look pretty. Now if you stuck your finger into the sand straight down from the top of the jar you would notice the sand has moved up the side of the jar to make space for your finger. The same as the atoms of the past would have to do for you. Now you take your finger out of the jar and the sand goes back to its original level. You look at it again and you say, how pretty. The grains of sand, however, are not in the same places that they were before you put your finger into the jar. You won't notice the difference, but there is a difference. That is what I think happened in my case. When you exit the past, there will be subtle changes that can in effect change the present.

There is another school of thought when talking about the space-time continuum, and that has to do with what I said about time being a ribbon and the speed at which time travels. You need to change the word from ribbon to track—as in train track—to understand it better.

In this theory, you have to think about the original idea of the ribbon but also that the single ribbon of time idea doesn't really work. People generally think of time travel as being a vast void that someone can enter and visit. Going where ever you want. But that can't be true.

The universe is for the most part an orderly place, with rules regarding how to engage it. Break the rules and utter destruction happens.

It is no different then driving along on a super highway and deciding to take the next exit. If you are in the left lane, you have to maneuver over to the right lane before taking an exit. You need to make sure that your path is clear or an accident may occur. If you do have an accident, it might or might not be catastrophic to you and your immediate surroundings, but it can and will effect the traffic miles away in all directions. This is how it works for the space-time continuum as well. In essence, there must be rules or natural laws to stop this from happening.

The rule as I see it is that time runs on tracks, many tracks, all starting from the same focal point, the Big Bang! You were born on a certain day on a track of time. I was born on another day on a different track of time. In this scenario, I may never meet you in the past or future because you are on your own track and I am on mine. On my track I can go backward into time and see what is in the past on my track, but it will be a different past than if you went back on your track, because they are separate tracks and never touch each other.

Well, just think about it for a minute! Think about how many things occur each and every day that you are not even aware of. How can you go back into time to see what you never knew about? It would be a new experience for you, and since you were not aware that it happened, how would you know how to react without knowing how it ended? You would most assuredly change the past and the space-time continuum, if it could be changed.

What I can do is travel up and down my track into the past and then into my new future as we discussed earlier in the book. During my life I have met certain people and have interacted with time, and it is that time that I can experience with time travel. You can experience your own events on your personal track of time but not my events.

Do you remember the world globes that used to be in the classrooms at school? I envision time tracks sort of like the longitude lines on the Earth globe. The lines that extend from the top to the bottom are the tracks of time, except that there are many more of them—so many that you can't even see a space in between them. They are not time itself; they are the tracks of time on which time will travel. This is where the speed of time comes in.

We have the Big Bang at the North Pole of the globe. Boom! Time starts, but you will notice that some of the tracks of time are moving faster then others. This is because time moves forward as each segment is filled. If a lot of things are happening, the segment fills up and time progresses along the track. If, on the other hand, there is a place void of things happening, time will slow down and wait for the segment to fill before advancing. This is true of the entire universe. I offer this statement not as proof but rather a theory that time doesn't move at a constant rate of speed—how could it?

What is the speed of time? What do you think? Are you moving right now? The answer is yes! Even if you are sitting in the most comfortable chair you have sipping tea, you are moving at a great speed because your chair is on the planet Earth, and the Earth is rotating on its axis at about 1,000 miles per hour. Not only is it rotating, but it is also traveling through space around the sun at 67,000 miles per hour. Then there is the sun traveling through space at 150 miles per second, and just think—we are moving right along with it. Then of course we have the Milky Way galaxy in which our sun travels. The Milky Way galaxy is also moving through the universe at a speed of 370 miles per second. Wow! That means you are not sitting in your chair (or standing still), you are moving through time with me at 370 miles per second. If our concept of the round universe is correct, that would mean it is also moving through whatever it would be called—the Super Universe, maybe—at who knows what speed.

So what is the speed at which time travels? I don't think that anyone will ever know because of the speed of everything else. I believe that time travels at a speed relative to where you are at that place and moment, which brings me to an important point: The speed of a place and the speed of time are closely intertwined. I believe that because of this very reason time travel is possible. There are places on your personal time track where time is moving very fast. And just like on a train moving at a high rate of speed, you can't just jump on and off of it easily. You can jump on and off a slow-moving train with little effort on your part.

As we have seen, time and speed are intertwined to such a degree that time itself moves at different rates of speed. Some places time moves fast; other places it moves slow. It is in this slow-moving mode that we will want to jump time and traverse it.

How do we know when time is moving slow? Just as light changes color at different speeds, and sound changes frequencies at different speeds, the lower the frequency the slower time is traveling. You simply set your time machine for the highest energy, lowest frequency that you can obtain. All that is required is that you match the electronic resonance of time and voila! You're in the past.

CHAPTER 8

Different Dimensions And Parallel Universes

Theoretically speaking, when you delve into quantum physics, parallel universes are supposed to exist, as are black holes. Black holes are now proven to exist, and with the space telescopes of today, we can even see them and also have photographs of them. New evidence now shows that indeed there are more than one in our own galaxy.

At the present time, the existence of parallel universes is just a theory based on mathematics. The problems arise because when someone comes up with a new formula and theory, the previously accepted theory gets put on a lower shelf and the new theory gets debated, dissected, and torn apart within the scientific community until it is either rejected or adopted. If rejected, the old theory resumes its former place as being correct.

Usually the decision to reject is based on some tiny flaw that cannot be explained, so the new theory might be correct, but because it has a flaw in its mathematics, it gets shelved.

Einstein once came up with a theory about time travel, but rejected the idea himself when the mathematics involved wouldn't prove out to his satisfaction. Years later, a physicist named Kurt Godel, building on Einstein's theory, came up with the correct answers to make Einstein's mathematics work. Mr. Godel's theory and ideas were rejected, though, because in order to make Einstein's mathematics work, it involved having a spiral universe, which was rotating, just as our galaxy, stars, and planets rotate. (And we all know that the universe is flat, right?)

The scientific community was not ready to except the idea of a spiral universe or that it was spinning. So his theory was put on a dusty shelf and forgotten.

What is a parallel universe? Supposedly it is a universe that is a twin to our own. The popular belief is that these parallel universes coexist, side by side, occupying the same space but in different dimensions. There would be a you and me in this other universe, and everything would be a twin of our universe, except not all things would be exactly the same. Maybe a light switch would be on the right side of the door, instead of on the left side. Maybe your eyes would be blue instead of brown. Subtle differences that you would not perceivably notice. I don't believe in the twin universe theory because of the word dimension! Do you know what a dimension is?

We live in a three-dimensional world. We see length, height, and depth. This dimension is very comfortable to us because it's our world. Take any one of those elements away and we would be lost and confused. This twin universe cannot be a three-dimensional universe because we are here living in our own three-dimensional universe, and two pieces of matter cannot occupy the same space. So this other universe would have to be a four-dimensional universe or an anti-matter universe, and if that were the case, it could not be a twin.

Try to visualize what dimensions are and you will see what I mean.

Picture yourself on top of a huge horizontal piece of ultra-smooth, flat clear plastic. It extends away from you for infinity, as far as you can see in front of you, as far as you can see on the sides and behind you. This is the first dimension.

Now a rod appears and the rod is as tall as we can see, and it goes down as far as we can see and is very ultra thin. It has to be thin because we cannot see depth yet. So now you can see height and width. This is the second dimension.

We now add depth and we can see the side of the rod. The third dimension.

To see the fourth dimension, let's use an empty shoebox to illustrate. We can look at the box and see length, width, and height. Now if we were in the fourth dimension, we not only see the length, width, and height, but we would also see the inside of the box and, I believe, the back. I really can't imagine a scarier place. You would have no reference to anything. If you happen to enter by mistake, you would never return.

A simple example: You enter the fourth dimension. You see a door. The doorknob is on the right side of the door, but because you are in the fourth dimension, you see the other side of the door also. It looks to you like a door with two knobs on it. Which knob do you turn? Are you exiting or entering through the door? Upon passing through the door, are you on the outside or the inside? You see both. Which is the your true position?

What I've told you is very simplified. Try imagining everything in our world in four dimensions. Do you think you could function in a four-dimensional world? It is a place I think I would not like to visit.

As far as anti-matter is concerned, it is said that if matter and anti-matter come into contact, they annihilate each other. Not a good idea to go there.

CHAPTER 9
How To Build a Time Machine

I have included a schematic and drawings of the time machine that I have designed to let you see what it looks like.

First, there is an outer shell that is stationary, and a framework attaches to it to hold it in place while it is running. The framework consists of four I-beams with cross sections in an A-frame style.

The outer shell I mentioned above is in the shape of a rectangle. It is made from a solid block of cobalt with the core drilled out to form a round cylinder twelve feet long and six feet in diameter. It is stationary and is part of the overall base of the machine. If you look at the diagram, you can readily see that the I-beams hold the entire weight of the cylinder. The working cylinders inside are free to move.

The entrance to the time machine is up the stairs on the right.

There are two moving cylinders: the outer cylinder and the inner cylinder.

The inside wall of the fixed-base core is lined with one-inch square magnetic tiles with the positive side of the magnets facing the closest and larger of the inner cylinders.

The outer wall of the largest inner cylinder is lined with half-inch magnetic tiles with the positive side facing the outer wall. This arrangement holds the inner cylinder in suspension.

There are what I call clamping bars bolted to the back and front ends of the fixed-base. They are in the shape of a cross hair and extend from the top of the base to the inner cylinder outside wall. There are grooves cut into the bars where the cylinder walls would touch. Since both inner cylinders are floating the cross bars are needed to stop the cylinders from floating away. The grooves allow the inner cylinders to turn while holding them from flying out of the time machine.

There also are 12 rubber shall we call them sleeves or rollers which are used as wheels to help control the spinning of the first cylinder closest to the base. They are mounted at equal distances along the base wall and rub against the outside wall of the larger inner cylinder. Four are at the front entrance to the machine four in the middle and four at the rear where the motor is located. The rear bottom right sleeve is the only one that has a shaft connected to it and it protrudes towards the rear of the machine. The shaft is about 3 inches in diameter and about 4-6 feet long. The size all depends on how large the motor is. The shaft is attached to a variable speed electric motor capable of reaching an undetermined amount of revolutions per minute. I am not sure if the shaft should be connected to the center of the motor, or if you should use a large gear to turn the shaft. That is part of the trial and error process. One way will work better and it is your fob to find out which. The motor and shaft must turn smoothly without the slightest vibration, as it is what rotates the outer cylinder so that you can start your trip.

The out side of the inner cylinder is also lined with half-inch magnetic tiles with the positive side facing out, which is towards the negative side of the magnets already installed. The resistance is tremendous, and you will find it is very hard to place the outer cylinder in the base. The same holds true for placing the inner cylinder into the outer cylinder. It will take a crew and some sort of vise like apparatus to hold them into place while you are attaching the bars to the ends of the base.

When power is applied to the motor, the outer cylinder rotates in a counter clockwise motion because the rubber sleeve or wheel pressing against the otter walls side is attached to the shaft which is in turn is connected to the center shaft of the motor which is turning clockwise. This turning creates a magnetic flux, which in turn should create a clockwise rotation of the inner cylinder. As the revolutions increase, the magnetic poles of the magnets start to reverse. When they do reverse, that will be the moment at which time travel is achieved.

You must also have the time traveler's suit. You will either have to have it made by a company that already is making flight suits or do it yourself.

But you must use electro-conductive fibers in place of the normal threads. Think of them as electric thread where a polymer fiber is given a metalized coating. If given a negative charge, then you will be floating inside of the inner cylinder and won't feel the effects of gravity.

That is how you will be transported to the past. These threads are ideal for electromagnetic interference shielding. They do exist now and are being developed. By the time you finish your time machine they may have made more progress and maybe a great breakthrough into this field. As you read further in this book towards the end, I tell you how to make the suit. I also offer a source for you to get the material in which to make it.

The only other apparatuses needed are the retaining straps that attach to your ankles and wrists. They are just fiberglass loops that slide over your ankles and wrists and are attached to a bigger one that fits around your torso. The size differs from person to person because of individual body structures. They are just there to stop your arms and legs from moving around and maybe scraping the walls of the tube. I chose fiberglass because it was nonmetallic, but anything will work—use your imagination. Heck, even a bungee cord will work. Just be sure to use plastic clips or hooks on the ends. No metal allowed!

Well, I guess that's it!

I have tried to explain to you in these pages what I feel and what I think can be accomplished. It is in my own way a feeble attempt to create in you a challenge that you will finish for me.

Good luck and happy landings. And may God be with you when- and where-ever you go.

THE END
OR IS IT?

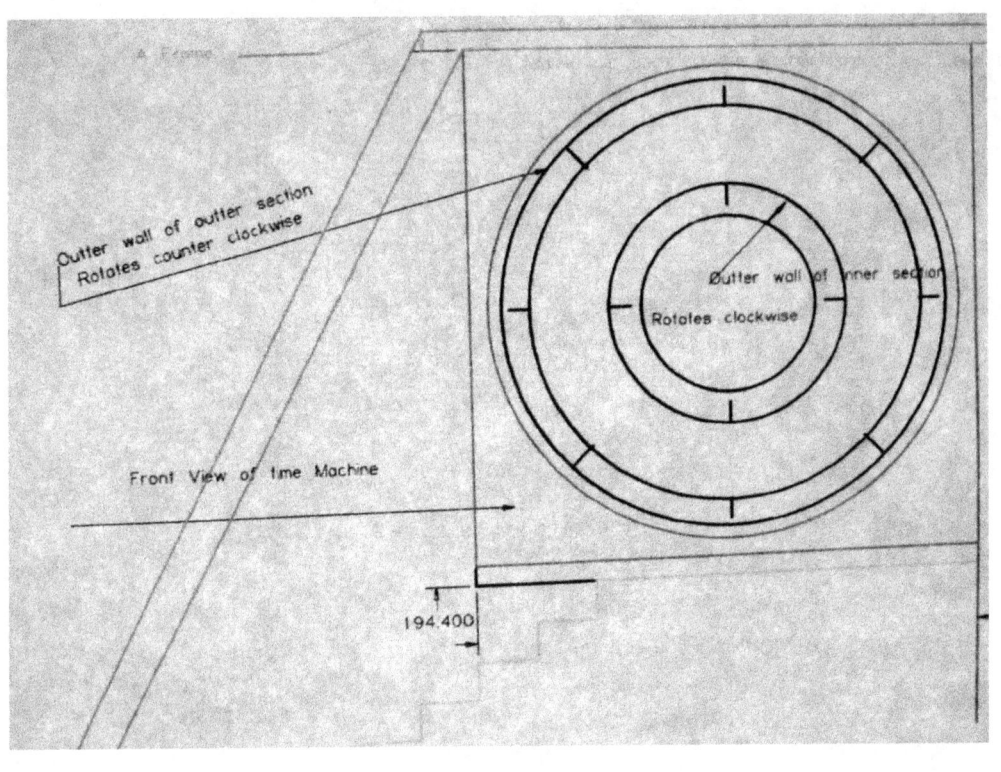

A. Frame

Outter wall of outter section
Rotates counter clockwise

Outter wall of inner section

Rotates clockwise

Front View of time Machine

194.400

Anthony J. Storino

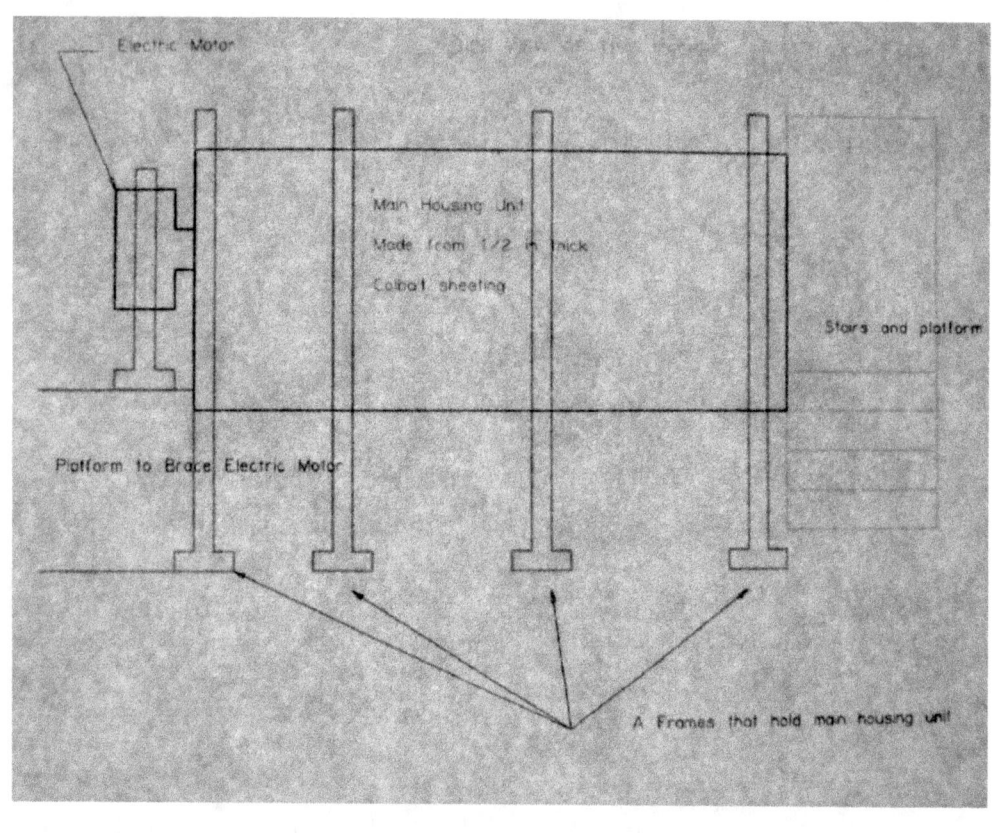

56

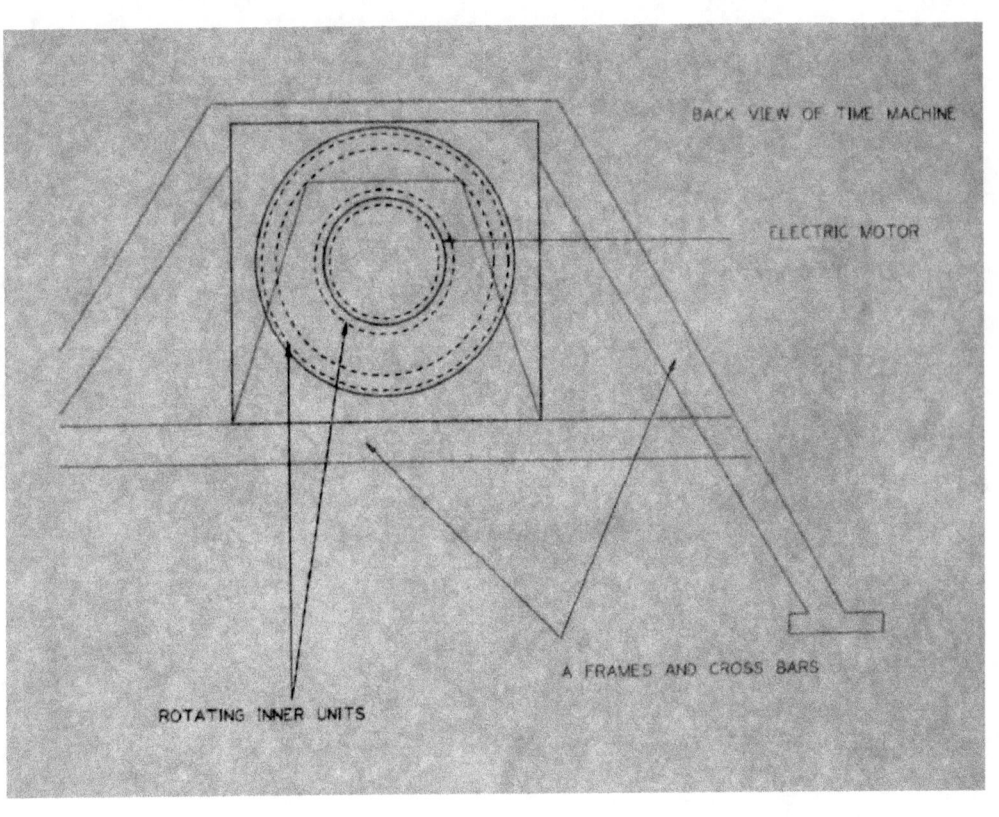

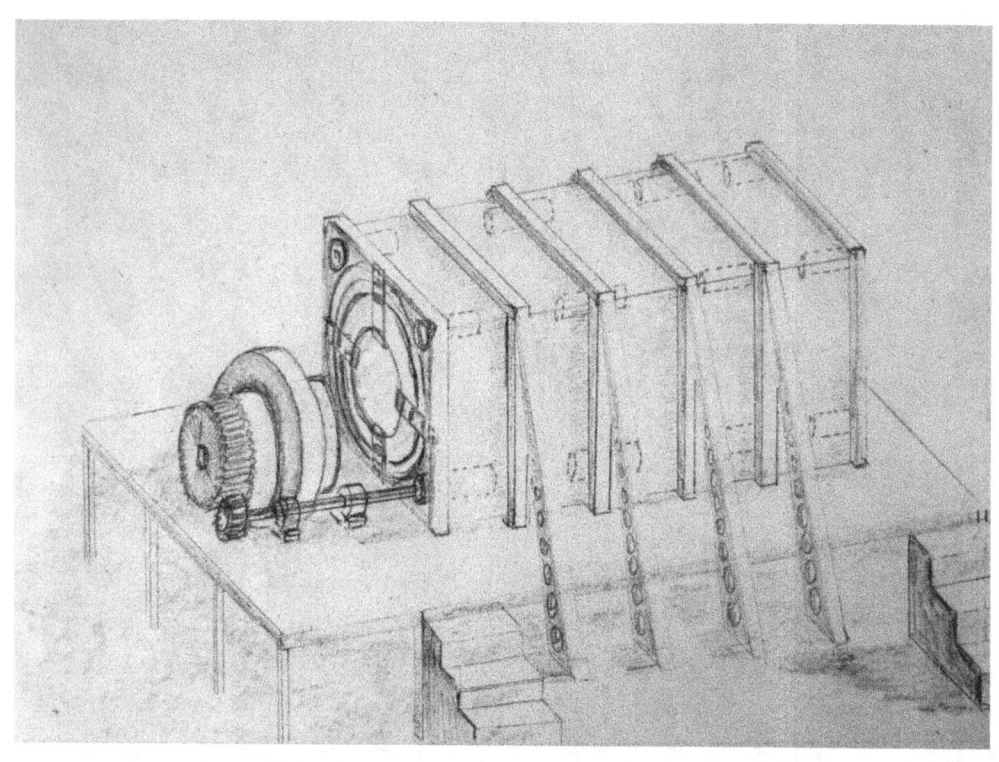

How To Make The Time Machine Suit

The suit is woven with the a product called AmberStrand. It is a flexible polymer/metal hybrid yarn. I believe it is made by Syscom Technology, Inc. If they are not the makers I am pretty sure they can tell you where to get it.

The suit uses Velcro sewn along the outside of the two flaps that have nylon zippers sewn in at their edges. The Velcro folds over the zippers to help keep any outside contaminants away from the user.

There are two zippers.

The first zipper starts on the outside of the leg at around the middle of the left thigh comes up across the waist and chest and ends just above the right nipple. The other zipper starts at the same height as the first at about the same place on the opposite leg. Follows the same path across the waist and chest and ends about the same area but at the opposite nipple. The suit is designed so both legs can be pushed in the suit easily before the zippers are closed. This is a one-piece suit with built-in shoes. I've toyed around with this idea, and for a built-in shoe to be comfortable, I think you should use a real shoe. Maybe a pair of your own shoes which you know are comfortable.

Then you could sew them into the suit so you would be assured of a good fit. The shoes must also be woven from the same material as the suit, but if you are going to use your own shoes I think you could somehow weave or sew the material around and attach it to your shoes. The material will have to surround your shoe, no gaps.

After you have your legs in, you stand up straight and pull the top of the suit over your head by sliding your arms into the sleeves and stretching your arms forward so the suit fits tightly over them. There is a hole at the top of the suit that is made from some sort of stretch material so it will fit tight around your neck. That neck part doesn't need to be made of the same suit material because your helmet will cover that part.

Pull the zipper up over your chest to your right nipple. Then pull the other zipper across your chest where a flap of material gets tucked in as you continue to your left nipple. Fold the Velcro down flat against the suit, overlapping the outside zipper.

The suit also has a set of gloves. I will leave the design up to you. They can be of any length just as long as they reach the forearm when worn. They are also made from the same material as the suit, must fit snug and must be worn while in the time machine. It would not be practical to have them sewn into the suit because you may need added dexterity in your fingers to do little things before you enter the time machine. If you think you can make a pair that can, say, pick up a dime, than by all means build them into the suit. Safety first is always best.

The helmet is a different story. It is made in layers using a soft inner lining for comfort with a high-impact gel coat layer on top of that. Once you arrive at the shape you desire, a layer of a closely webbed suit-material in the form of a fiber lattice should be applied. Followed by another layer in the opposite direction. Then another coating of high-impact gel coat gets placed on top of that. Shape the outside of the helmet to your desire.

The visor is made from a thin cobalt/metal screen sandwiched between two pieces of Plexiglas and shaped to the opening on the helmet. Gel coat it to the helmet, sand it, and finish it any way you like.

If you look on the Internet, you can find an assortment of used high-altitude flying helmets or surplus space helmets. You could conceivably buy one that fits nicely and then apply your mesh work on top of it, and that way you would already have a full face visor to work with.

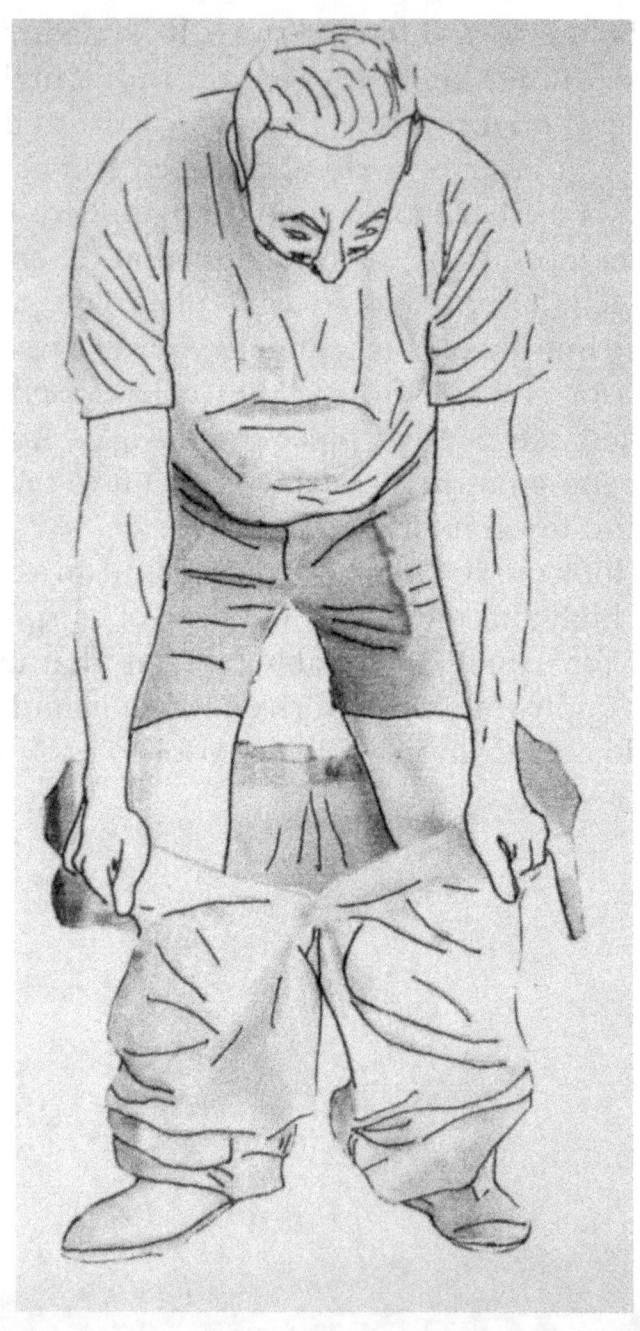

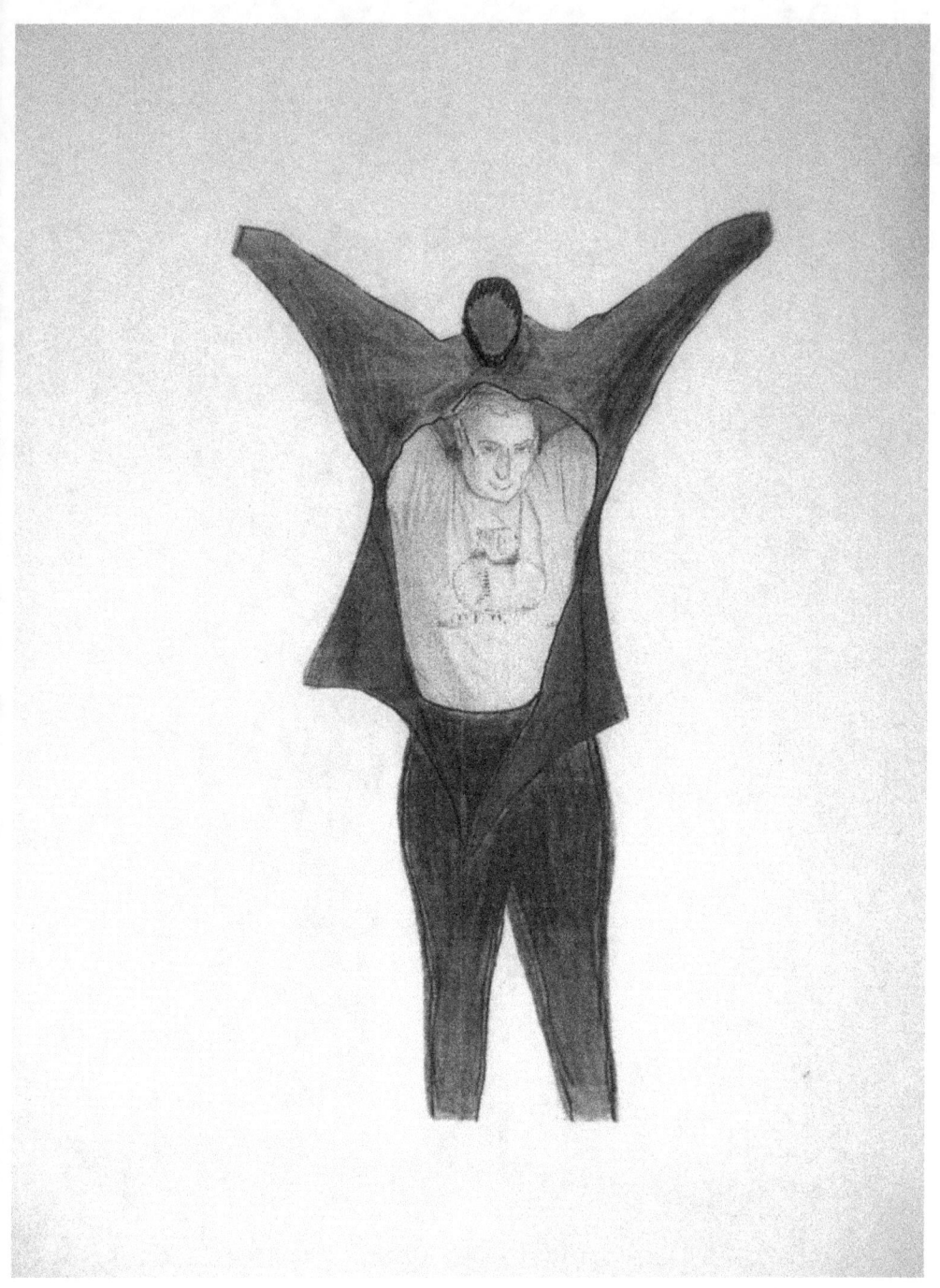

63

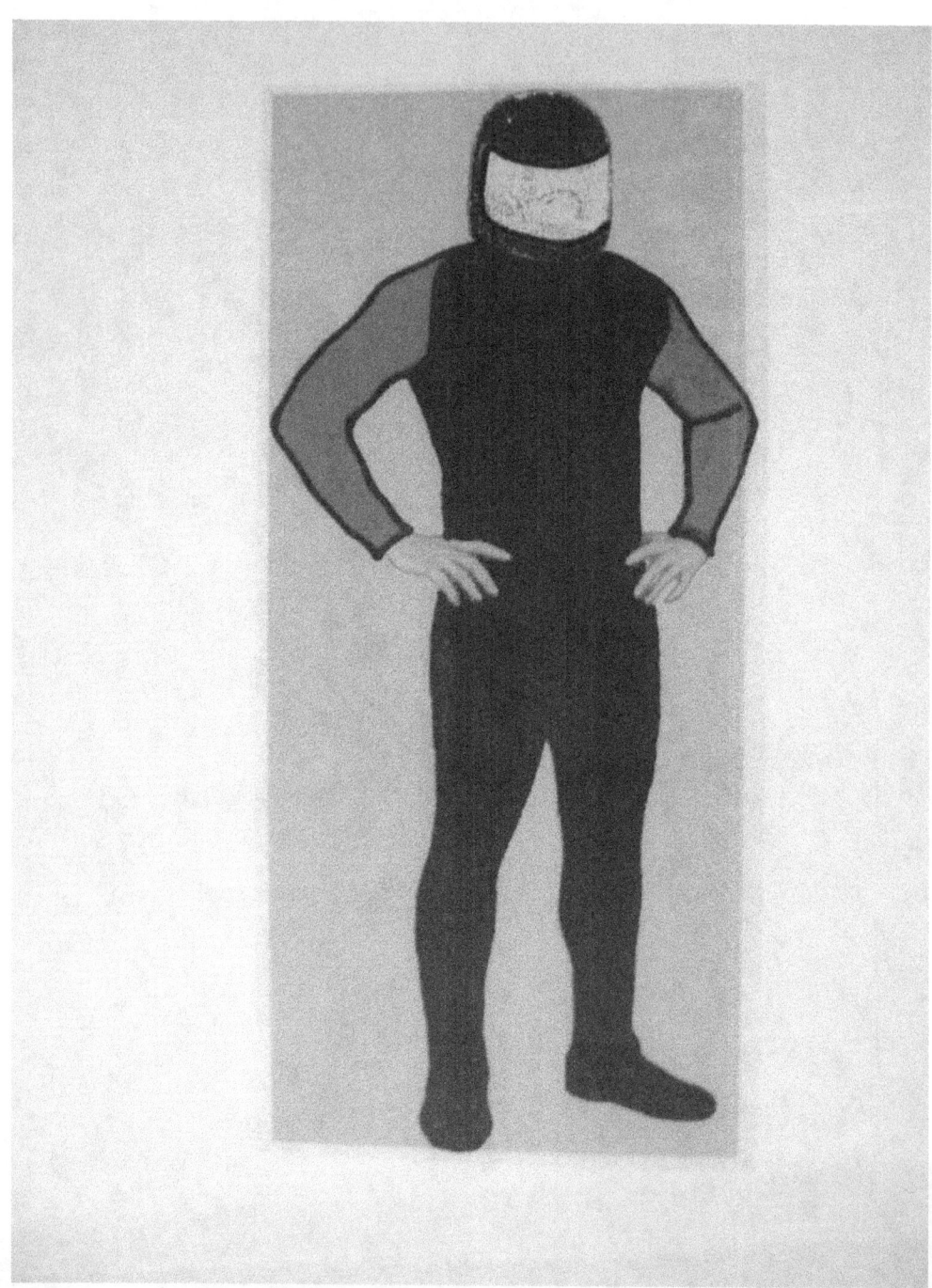

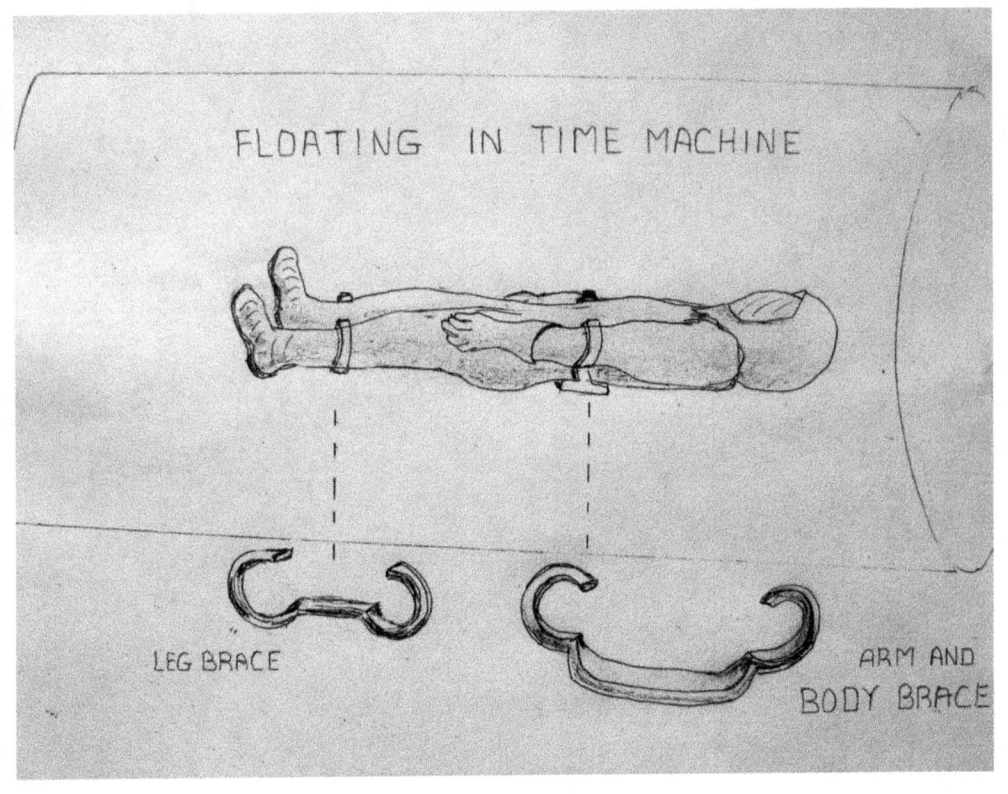

Conclusion

Well now you have it! My ideas based on what I have learned in my lifetime. Throughout my lifetime I have read many articles connected with this subject so that they sort of intertwine in my brain. I come up with a plan or idea of how a time machine should work, then I find some new data, or old data that is new to me and the old light bulb once again turns on, and a new image starts to form in my brain. I work on it change it, mold it into something that might just work if I only had this! Then I start looking for the missing piece. During the refining process I find errors in what I believed was my final outcome and start to pursue new avenues of approach. This in turn sometimes not always changes my thoughts on other things and processes that have nothing to do with what I was refining.

Questions, questions and more questions! You need the questions or you will be stagnant in your work. Only through constant probing and giving way to new ideas and thoughts can you ever hope to finalize your ideas on this or any other subject.

I leave it up to you, the curious, the knowledgeable, the ones with the desire to find out the answers to some of the puzzles that have plagued man from the start of his short existence here on earth.

Question and learn from your research. Come away forming your own ideas and if need be change or throw out the accepted doctrine of the day. You will be scaling new heights and forging new trails into the unknown. Take what I offer, and if it displeases you, change it, mold it into your own creation. Find the missing piece of the puzzle and if it doesn't fit, change its shape a little until you can make it fit in with your new ideas.

The only one that can stop you is you.

The End

Extra Books To Read

I would suggest reading a book by Fred Alan Wolf entitled Parallel Universes. It is easy reading and flows smoothly.

The other is for those of you who would like to know more about the life of Nikola Tesla.

It is simply called Tesla and is written by Tad Wise. I have many books about Tesla, but they tell of his inventions more than about the man himself. I think you will enjoy it.